Uranus

The Tilted Rebel

JD ARDEN

Uranus

Preface: A Quiet Outlier

Uranus is a planet of contradictions. It is both immense and subtle, a giant world cloaked in a pale blue haze that masks its inner complexities. Alone among the planets in the solar system, Uranus rolls along its orbit, its axis tipped at an extreme angle that defies convention. It is a place of extremes—icy temperatures, faint rings, and moons named after literary characters—yet it often seems overshadowed by its brighter, stormier neighbor, Neptune, or the grand gas giants closer to the Sun.

To dismiss Uranus as a quiet, uneventful planet, however, would be a mistake. Beneath its understated appearance lies a world of profound mysteries and surprises. Its peculiar tilt, which gives it seasons unlike any other planet, challenges our understanding of planetary formation and dynamics. Its magnetic field, wildly off-center and tilted at an angle, offers a glimpse into the hidden processes of its interior. Its rings, faint and narrow, reveal a unique dynamism, while its moons, each with its own quirks, tell stories of geological activity and evolution.

In many ways, Uranus is a rebel—a planet that refuses to conform to expectations. It invites us to look beyond the obvious, to seek out the subtle and the unusual, and to appreciate the diversity of planetary systems. Uranus is not a planet that demands attention with dramatic storms or dazzling rings; it rewards those who look deeper, who ask questions about the unseen forces that shape worlds.

This book explores Uranus as a symbol of quiet revolution and understated complexity. From its unique tilt and magnetic field to its icy atmosphere and family of moons, Uranus challenges us to expand our understanding of what a planet can be. It is a reminder that even in the cold, distant reaches of the solar system, there are wonders waiting to be discovered, and that the value of difference lies not in how it compares to the familiar, but in how it reveals new possibilities.

Chapter 1: A World on Its Side

Uranus stands out among the planets for one startling reason: it is tipped on its side. Unlike every other major planet in the solar system, which rotates more or less upright relative to their orbits, Uranus's axis is tilted by 98 degrees. This extreme orientation means that Uranus essentially rolls along its orbit, with one pole pointing almost directly at the Sun for part of its 84-year orbital period, before the other takes its turn. This unique tilt creates some of the most bizarre seasonal patterns in the solar system and raises profound questions about the planet's history and internal dynamics.

The axial tilt of Uranus is thought to be the result of a colossal collision early in the planet's history. During the chaotic period of planetary formation, Uranus may have been struck by a massive protoplanet, one large enough to knock it onto its side. This impact would have reoriented the planet's axis, altered its rotation, and perhaps even affected its internal structure and atmosphere. While the exact details of this event remain speculative, the tilt is a visible scar, a reminder of the violent processes that shaped the solar system.

The consequences of this tilt are profound. On Earth, the axial tilt of 23.5 degrees gives rise to the familiar progression of seasons, as different parts of the planet receive varying amounts of sunlight throughout the year. On Uranus, the extreme tilt creates seasons that are not only more dramatic but also vastly longer. Each pole experiences 42 years of continuous sunlight followed by 42 years of darkness, while the equatorial regions experience a more chaotic interplay of light and shadow.

The transition between seasons on Uranus is marked by unusual atmospheric changes. During its equinoxes, when the Sun shines directly over the planet's equator, Uranus's atmosphere appears to become more active. Observations from Earth-based telescopes and the Hubble Space Telescope have revealed occasional storms and bright spots in the otherwise featureless haze of Uranus's clouds. These changes suggest that the planet's unique orientation affects the circulation patterns and dynamics of its atmosphere in ways that are not yet fully understood.

Uranus

The tilt of Uranus also influences its internal dynamics. The planet's rotation, with its axis almost parallel to its orbital plane, creates unusual interactions between its magnetic field and the solar wind. Unlike Earth's magnetic field, which is roughly aligned with its rotational axis, Uranus's magnetic field is tilted by an additional 59 degrees and is significantly off-center. This misalignment creates a magnetic environment that is both chaotic and unique, with charged particles spiraling along paths that defy the orderly patterns seen around other planets.

This combination of extreme axial tilt and magnetic irregularities makes Uranus a fascinating case study for planetary scientists. It challenges traditional models of how planets form and evolve, suggesting that chance events like collisions can have lasting, transformative effects. Uranus's tilt also invites comparisons to exoplanets, many of which are thought to have undergone similarly dramatic interactions during their formation. By studying Uranus, scientists gain insights into the diversity of planetary systems and the forces that shape them.

Philosophically, Uranus's tilt is a metaphor for the value of difference. It is a planet that refuses to conform to the norms of its peers, a world that challenges our assumptions about what is normal or expected. Its unique orientation reminds us that the cosmos is filled with diversity, and that the most interesting and illuminating phenomena often arise from the unexpected.

To observe Uranus in its quiet, rolling orbit is to witness a planet that embodies resilience and adaptation. It has survived the chaos of its early history, retained a family of moons and rings, and continues to orbit the Sun with quiet dignity. Its seasons may be extreme, its dynamics complex, but Uranus remains a testament to the variety and ingenuity of nature's designs.

In the end, Uranus's tilt is not a flaw but a feature—a reminder that even in the vastness of space, uniqueness has its place, and that the most profound discoveries often come from worlds that challenge us to see differently.

Chapter 2: The Pale Blue Mystery

Uranus's appearance is one of quiet simplicity: a serene, pale blue orb that drifts through the outer reaches of the solar system. At first glance, its understated hue and lack of prominent features might suggest a planet of monotony. Yet, this unassuming exterior masks a world of profound mysteries, one that challenges scientists to decipher its secrets. Unlike the dynamic storms of Jupiter or the deep azure turbulence of Neptune, Uranus's pale blue facade presents a puzzle: Why does this planet appear so subdued, and what lies beneath its tranquil exterior?

The distinctive color of Uranus is a result of its **atmospheric composition**. Like other gas and ice giants, Uranus's atmosphere is composed predominantly of hydrogen and helium, with smaller amounts of methane. It is the methane that gives Uranus its blue-green hue; this gas absorbs red wavelengths of sunlight while reflecting blue and green light back into space. This same process is responsible for the color of its sibling, Neptune, but the contrast between the two is striking. Neptune's deep, rich blue is far more vibrant than Uranus's muted tones, a difference that has intrigued planetary scientists for decades.

The mystery of Uranus's pale appearance lies not only in its color but also in its lack of visible atmospheric activity. While Neptune frequently exhibits dark spots and bright storms—some as large as Earth itself—Uranus appears almost featureless. Early observations by telescopes and the **Voyager 2 spacecraft**, which flew by Uranus in 1986, revealed only subtle bands and faint, transient storms. Compared to its neighbor, Uranus seemed eerily calm, an enigma in a solar system filled with dynamic worlds.

The apparent tranquility of Uranus is likely tied to its **internal heat—or lack thereof**. Unlike Neptune, which radiates more than twice the energy it receives from the Sun, Uranus emits almost no excess heat. This lack of internal warmth is a significant anomaly. Most planets, especially gas and ice giants, retain heat from their formation or generate it through gravitational compression, radioactive decay, or chemical processes. Uranus, however, seems to have lost—or never developed—a mechanism to retain or release significant internal energy.

Uranus

This absence of internal heat has profound implications for Uranus's atmosphere. Without the energy to drive convection and circulation on a large scale, the planet's weather systems are muted. Clouds form and move at a slower pace, and storms are rare and short-lived compared to those on other planets. The result is an atmosphere that, at least superficially, appears calm and unchanging.

The origins of Uranus's peculiar energy deficit remain a subject of debate. One hypothesis suggests that the same cataclysmic impact that tipped Uranus onto its side may also have disrupted its core, scattering heat-generating materials or creating layers that inhibit the flow of energy. Another theory posits that Uranus's atmosphere contains components that trap heat more effectively than previously thought, preventing it from escaping into space. Regardless of the cause, this energy deficit is a key factor in shaping Uranus's unique character.

Despite its subdued appearance, Uranus is not entirely devoid of activity. Observations in recent decades, particularly with the Hubble Space Telescope and ground-based telescopes equipped with advanced optics, have revealed fleeting signs of atmospheric dynamics. Bright spots and faint storms occasionally appear, often during the planet's equinox, when the Sun shines directly on its equator. These events suggest that Uranus's atmosphere is more dynamic than it appears, with processes that may be tied to its extreme axial tilt and seasonal changes.

The **vertical structure** of Uranus's atmosphere adds another layer of intrigue. The uppermost layer, called the troposphere, is where most weather phenomena occur. Beneath it lies the stratosphere, a region where temperatures rise due to the absorption of ultraviolet radiation by methane and other gases. Farther below, the pressure increases dramatically, compressing gases into liquid or supercritical states. The transition from these upper atmospheric layers to the dense interior is poorly understood, but it is thought to involve exotic forms of matter, including water, ammonia, and methane compressed into icy or slushy states.

These deeper layers may also hold the key to understanding Uranus's pale color. One possibility is that hazes or aerosols in the atmosphere scatter light differently than those on Neptune, muting the planet's hue. Another theory suggests that Uranus's deeper clouds, which may contain hydrogen sulfide or other compounds, interact with sunlight in ways that produce its distinctive blue-green tones. Understanding these processes requires further exploration and more detailed observations, both from Earth and future spacecraft.

Uranus's muted appearance also raises broader questions about how planets develop and evolve. Its calm exterior, energy deficit, and unique tilt make it a valuable counterpoint to Neptune, offering insights into the diversity of ice giants and the factors that shape their atmospheres. By comparing Uranus to its neighbors, scientists can refine models of planetary formation and behavior, extending those insights to the growing catalog of exoplanets in other star systems.

Philosophically, Uranus's pale blue mystery is a reminder that appearances can be deceiving. What seems tranquil and unremarkable on the surface often hides complexity and depth. Uranus challenges us to look beyond the obvious, to ask questions about what lies beneath, and to appreciate the subtleties that define its character.

To observe Uranus is to see a world that invites curiosity. Its understated beauty, its serene hue, and its quiet demeanor stand in stark contrast to the more dramatic worlds of the solar system. Yet, in its quiet rebellion against expectations, Uranus reveals a different kind of wonder—a planet that does not shout its mysteries but whispers them, waiting for those willing to listen.

Chapter 3: Uranus's Magnetic Dance

Uranus's magnetic field is as unconventional as the planet itself. It is neither aligned with the planet's rotation nor centered within the planet's core. Instead, Uranus's magnetic field is wildly tilted—offset by 59 degrees from the rotational axis—and significantly off-center, with its origin lying far from the planetary core. This unusual configuration creates a magnetic environment that is dynamic, chaotic, and unlike any other in the solar system.

Most planetary magnetic fields, including Earth's, are generated by the motion of conductive fluids in the core, a process known as the **dynamo effect**. These fields are typically aligned closely with the planet's rotational axis, creating a stable and predictable relationship between the planet's spin and its magnetic poles. Uranus, however, defies this norm. Its magnetic poles are located far from its geographic poles, and the field is asymmetric, varying significantly in strength and structure as it rotates.

The origins of Uranus's peculiar magnetic field are closely tied to its unique history. Scientists believe that the same massive collision that tipped Uranus onto its side may also have disrupted the internal layers responsible for generating the magnetic field. Unlike Earth or Jupiter, where the dynamo effect occurs deep in the core, Uranus's magnetic field may be generated in a thin shell of electrically conductive material closer to its outer layers. This shell could consist of a mixture of water, ammonia, and methane compressed into an exotic state called **superionic ice**, where ions flow freely, creating the conditions needed for a dynamo.

The result of this unusual configuration is a magnetic field that behaves unpredictably. As Uranus rotates on its tilted axis, the magnetic field wobbles and shifts, creating regions of varying intensity and complexity. At certain points in Uranus's rotation, the field may be strong and well-defined, while at others, it may weaken or become distorted. This irregularity creates a magnetic environment that is far more dynamic than the relatively stable fields of Earth or Jupiter.

Uranus

One of the most striking consequences of Uranus's magnetic field is its effect on the planet's **magnetosphere**, the region of space dominated by the planet's magnetic influence. Uranus's magnetosphere is not centered on the planet and does not align neatly with its rotation, resulting in a lopsided and constantly shifting structure. The magnetosphere is compressed on the side facing the Sun and stretched into a long tail on the opposite side, much like Earth's, but the tilt and offset of Uranus's magnetic field create a more complex and irregular shape.

This dynamic magnetosphere interacts with the solar wind—the stream of charged particles emitted by the Sun—in unique ways. As Uranus orbits the Sun, its rotation causes the magnetosphere to sweep through space like a wobbling top, creating waves and eddies in the charged particles surrounding the planet. These interactions produce **auroras** near Uranus's magnetic poles, though they are far more irregular and diffuse than the auroras seen on Earth or Jupiter.

The unusual behavior of Uranus's magnetic field also has implications for its moons and rings. The charged particles trapped within the magnetosphere interact with these objects, creating subtle changes in their surfaces and influencing the dynamics of the rings. For example, the moons embedded within Uranus's magnetosphere may experience surface weathering due to the bombardment of charged particles, altering their appearance over time.

Studying Uranus's magnetic field offers insights not only into the planet itself but also into the broader processes that shape magnetic fields across the cosmos. Magnetic fields are a common feature of planets, stars, and even galaxies, and their behavior is influenced by the composition, structure, and dynamics of the objects that generate them. By examining Uranus's magnetic field, scientists can refine their understanding of how magnetic fields form, evolve, and interact with their environments.

Uranus's magnetic field also serves as a point of comparison for exoplanets—planets orbiting stars beyond our solar system. Many of these distant worlds are thought to be similar in size and composition to Uranus, and their magnetic fields may share some of its unusual characteristics. By studying Uranus, we can develop models that help interpret the observations of exoplanetary magnetic fields, offering clues about their internal structures and atmospheric dynamics.

Uranus

Philosophically, Uranus's magnetic field is a reminder of the unexpected complexity that lies beneath seemingly simple exteriors. The planet's pale blue facade gives no hint of the dynamic and chaotic forces at work within its magnetosphere, challenging us to look beyond appearances and consider the hidden processes that shape the universe.

Uranus's magnetic dance is a testament to the diversity of planetary systems and the creativity of nature's designs. It is a field that defies expectations, a force that wobbles and shifts, a reminder that even in the vastness of space, there are phenomena that surprise and intrigue. To study Uranus's magnetic field is to embrace the unknown and to marvel at the ingenuity of a universe that is as unpredictable as it is beautiful.

Chapter 4: The Rings of Uranus

The rings of Uranus, faint and understated compared to the dazzling splendor of Saturn's, are nonetheless a testament to the planet's quiet elegance and hidden complexity. They are a system of narrow, dark bands encircling the planet, composed primarily of ice and rock, with particles ranging in size from fine dust to larger boulders. First discovered in 1977, Uranus's rings have since been revealed to be dynamic and enigmatic, shaped by the interplay of gravity, collisions, and the subtle influence of Uranus's moons and magnetic field.

Unlike the bright, icy rings of Saturn that gleam in sunlight, Uranus's rings are dark, reflecting only a small fraction of the light that strikes them. This muted appearance is likely due to the composition of the ring particles, which are thought to be coated with dark, radiation-processed materials, such as organic compounds or carbonaceous dust. The exact origin of these particles remains uncertain, but their presence speaks to the diversity of processes at work in the Uranian system.

The ring system of Uranus is divided into 13 distinct rings, each with its own unique properties. The **epsilon ring**, the brightest and most prominent, lies at the outer edge of the system and is flanked by two faint, dusty rings. The inner rings are narrower and fainter still, creating a delicate structure that challenges observation and study. Between the main rings lie diffuse bands of fine particles, forming a faint halo that extends farther from the planet.

The narrowness and sharp edges of Uranus's rings are among their most intriguing features. Unlike the broad, sprawling rings of Saturn, Uranus's rings are confined to narrow bands, often only a few kilometers wide. This precision is maintained by the gravitational influence of **shepherd moons**, small satellites that orbit within or near the rings. These moons, such as Cordelia and Ophelia, exert a stabilizing effect on the ring particles, preventing them from spreading out and maintaining the sharp boundaries of the rings.

The origins of Uranus's rings remain a mystery, though several theories offer possible explanations. One hypothesis suggests that the rings are the remnants of a moon or moons that were shattered by a collision or torn apart by Uranus's tidal forces. Over time, the debris from this event could have coalesced into the narrow rings observed today. Another possibility is that the rings formed alongside Uranus during its early history, a byproduct of the same processes that shaped the planet and its moons.

Uranus's rings are not static; they are dynamic systems, constantly evolving under the influence of gravity, collisions, and the planet's magnetic field. Particles within the rings collide with one another, breaking apart and reaggregating in a delicate balance. The faint, dusty halo of particles that surrounds the main rings is evidence of this ongoing activity, as collisions generate smaller fragments that diffuse outward.

The discovery of Uranus's rings in 1977 was a milestone in planetary science. Astronomers James Elliot, Edward Dunham, and Douglas Mink first detected the rings while observing a star as it passed behind Uranus. They noticed that the star's light dimmed intermittently before and after it was obscured by the planet, revealing the presence of several rings. This discovery was later confirmed by the **Voyager 2 spacecraft**, which captured detailed images of the rings during its flyby in 1986. These observations transformed Uranus from a relatively obscure world into a planet of significant scientific interest.

The study of Uranus's rings provides valuable insights into the processes that govern planetary ring systems. By comparing Uranus's rings to those of Saturn, Jupiter, and Neptune, scientists can identify commonalities and differences that shed light on the factors that shape these structures. Uranus's rings, with their narrow bands and dark composition, represent a unique case, offering clues about the diversity of ring systems and the conditions under which they form.

The interaction between Uranus's rings and its moons is another area of fascination. The gravitational pull of the moons not only stabilizes the rings but also creates waves and ripples within them, adding to their complexity. These interactions highlight the interconnectedness of Uranus's system, where the motions of moons, rings, and the planet itself are inextricably linked.

Uranus

Philosophically, the rings of Uranus are a symbol of subtlety and restraint. They lack the grandeur of Saturn's rings, yet they possess a quiet elegance that rewards close observation. Their narrowness, darkness, and dynamic nature remind us that beauty in the cosmos is not always obvious or ostentatious; it often lies in the details, in the delicate balance of forces that sustain complex systems.

To observe Uranus's rings is to witness a system in motion, a structure that exists not in isolation but as part of a larger, dynamic environment. The rings are shaped by collisions and gravity, influenced by Uranus's moons and magnetic field, and subject to the inexorable forces of time. They are a reminder that even the most understated features of the universe are products of extraordinary processes, and that their beauty lies in their complexity and transience.

Uranus's rings, though faint and often overlooked, are a testament to the planet's quiet rebellion against expectations. They are a feature of subtlety and significance, a symbol of the diversity of planetary systems and the richness of the cosmos. To study them is to appreciate the intricate interplay of forces that creates harmony in the heavens and to marvel at the quiet elegance of a planet that refuses to conform.

Chapter 5: A Family of Moons

Uranus is home to a remarkable family of moons, each with its own story to tell. Unlike the moons of Jupiter and Saturn, which are often named after mythological gods and titans, the moons of Uranus take their names from the works of William Shakespeare and Alexander Pope, giving them a literary flair that complements the planet's unconventional nature. These satellites, ranging from small, irregular bodies to larger, geologically intriguing worlds, form a diverse and dynamic system that reflects Uranus's own unique character.

The Moons of Uranus: An Overview

To date, 27 moons have been confirmed to orbit Uranus, divided into three categories based on their size, orbits, and characteristics: **major moons**, **regular moons**, and **irregular moons**. The major moons—Titania, Oberon, Umbriel, Ariel, and Miranda—are the largest and most studied, each presenting fascinating geological features and histories. The smaller, regular moons orbit closer to Uranus and are often shepherds of its rings, stabilizing and sculpting their structure. The irregular moons, likely captured objects from the Kuiper Belt or beyond, follow distant, eccentric orbits and remain shrouded in mystery.

Titania: The Queen of the Moons

Titania, the largest moon of Uranus, is named after the queen of the fairies in Shakespeare's *A Midsummer Night's Dream*. With a diameter of 1,578 kilometers, it is roughly half the size of Earth's Moon but displays a surface marked by a complex interplay of craters, canyons, and plains. Evidence suggests that Titania has undergone significant geological activity in its past, with vast fault lines and chasms indicating tectonic processes.

One of the most striking features of Titania is its network of **graben**, or rift valleys, which likely formed as the moon's interior cooled and contracted. These features hint at a dynamic history, suggesting that Titania may have once had a subsurface ocean or a period of internal heating that shaped its surface. While no current evidence points to ongoing geological activity, Titania remains a candidate for further exploration, especially in the search for signs of ancient habitability.

Oberon: The Distant Companion

Oberon, named after Titania's consort in *A Midsummer Night's Dream*, is the second-largest moon of Uranus and the most distant of the major moons. Its heavily cratered surface suggests an ancient and static history, with little evidence of recent geological activity. However, dark material seen in some of its craters hints at the possibility of past eruptions or impacts that exposed subsurface material.

Despite its relative inactivity, Oberon's isolation and surface features make it an intriguing target for study. Its heavily cratered landscape serves as a record of the early solar system, preserving the impacts and processes that shaped Uranus's moons in their infancy.

Ariel: The Brightest and Most Dynamic

Ariel, the brightest of Uranus's major moons, named after the airy spirit in Shakespeare's *The Tempest*, displays a surface that is among the youngest and most geologically active in the Uranian system. Crisscrossed by canyons, ridges, and plains, Ariel's landscape suggests a history of tectonic activity and resurfacing events. Observations by the Voyager 2 spacecraft revealed bright icy deposits along canyon walls, indicating that Ariel may have experienced cryovolcanism, with water or other ices erupting from its interior.

Ariel's brightness is due to its relatively high albedo, reflecting a significant portion of the sunlight that strikes it. This, combined with its diverse terrain, makes Ariel a key target for understanding the processes that have shaped Uranus's moons and their potential for hosting subsurface oceans.

Umbriel: The Dark and Mysterious

Umbriel, named after a melancholic sprite in Alexander Pope's *The Rape of the Lock*, is the darkest of Uranus's major moons, with a surface that absorbs more sunlight than any of its siblings. Its gloomy appearance is matched by its geological quietness; Umbriel shows few signs of tectonic or volcanic activity, suggesting a long history of inactivity.

Despite its dark surface, Umbriel is not without intrigue. A prominent bright ring, nicknamed **Wunda Crater**, stands out against the moon's otherwise shadowy landscape. The origin of this feature is uncertain but could be the result of an impact that exposed brighter material beneath the surface. Umbriel's darkness may also provide clues about the composition and weathering of its surface over time.

Miranda: The Fractured World

Perhaps the most enigmatic of Uranus's moons, Miranda is a small, irregularly shaped world with a surface that defies easy explanation. Named after the tragic heroine of *The Tempest*, Miranda is a patchwork of craters, ridges, and cliffs, with features that appear mismatched and chaotic. Its most famous landmark, **Verona Rupes**, is a cliff that may be the tallest in the solar system, plunging nearly 20 kilometers from top to bottom.

Miranda's fractured surface suggests a history of dramatic geological upheaval, possibly caused by tidal forces exerted by Uranus. One theory posits that Miranda was shattered by a massive impact and reassembled from its own fragments, creating its jumbled appearance. Another hypothesis suggests that internal heating, perhaps driven by tidal interactions, reshaped its surface in multiple episodes.

Miranda's small size—only 470 kilometers in diameter—belies its complexity, making it a fascinating target for future exploration.

The Shepherds and Stragglers

Beyond the major moons, Uranus's smaller moons play vital roles in shaping its ring system. Moons like Cordelia, Ophelia, and Bianca act as shepherds, their gravity maintaining the narrowness and structure of Uranus's rings. These tiny satellites, often only tens of kilometers across, are difficult to study but essential to the dynamics of the Uranian system.

Farther out lie the irregular moons, such as Sycorax and Caliban, which follow eccentric orbits that suggest they were captured by Uranus's gravity rather than forming alongside it. These moons may be remnants of the Kuiper Belt or other primordial regions of the solar system, offering insights into the processes that shaped the outer planets and their satellites.

A Family of Stories

The moons of Uranus are more than geological curiosities; they are a tapestry of stories that reflect the planet's tumultuous history and dynamic present. From Titania's canyons to Miranda's fractured surface, each moon reveals a unique aspect of Uranus's character. They challenge us to expand our understanding of moons not as static bodies but as evolving worlds, shaped by forces both internal and external.

Philosophically, Uranus's moons symbolize diversity and resilience. They are worlds of contrast—bright and dark, active and dormant, smooth and fractured—that remind us of the complexity of planetary systems. Together, they form a family that is as unconventional and intriguing as Uranus itself, a testament to the variety and beauty of the cosmos.

Chapter 6: The Quiet Giant in Myth and Discovery

Uranus has always been an outlier, a planet that defies both expectation and easy categorization. It was the first planet discovered in modern times, a world hidden in plain sight until advances in technology and science brought it into focus. Unlike the planets known to ancient civilizations, which were visible to the naked eye and woven into myth and culture over millennia, Uranus remained unknown until the late 18th century. Its discovery marked a turning point in astronomy, reshaping humanity's understanding of the solar system and revealing that the heavens were far more expansive than previously imagined.

In 1781, the English astronomer William Herschel observed a faint object through his telescope that he initially believed to be a comet. Careful observation over subsequent nights revealed that it moved too slowly and in too orderly a path to be a comet. Herschel had stumbled upon something far more significant: the first new planet discovered since antiquity. This groundbreaking find extended the known boundaries of the solar system, nearly doubling its perceived size overnight. It was a revelation that underscored the power of observation and technology to uncover the unseen.

The discovery of Uranus was a pivotal moment not only for astronomy but also for the broader intellectual climate of the Enlightenment. It demonstrated that human curiosity and ingenuity could reveal truths about the cosmos that had eluded even the most advanced ancient thinkers. Uranus's existence challenged long-held beliefs about the limits of the universe and the static nature of the heavens. It became a symbol of progress, a planet whose discovery spoke to the potential of reason and the scientific method.

Yet, Uranus remained a quiet enigma even after its discovery. Unlike Jupiter and Saturn, whose striking appearances made them objects of wonder and mythology, Uranus presented no dazzling rings or colorful storms to captivate early observers. Its pale, featureless disk seemed almost indifferent to its newfound prominence. For decades, it lingered

Uranus

in the shadow of its more dynamic neighbors, its mysteries hidden beneath a layer of muted blue-green haze.

Culturally, Uranus did not immediately inspire the myths and legends that surrounded the other planets. Its name, chosen to align with classical mythology, paid homage to the primordial sky god of Greek mythology, Uranus, the father of Cronus (Saturn) and grandfather of Zeus (Jupiter). In this way, Uranus was linked to the existing mythological framework, completing a family of celestial deities. Yet the god Uranus was a relatively obscure figure, and the planet's remote and understated nature seemed to reflect this quiet lineage.

The lack of historical myths associated with Uranus allowed it to occupy a unique cultural space. It became a planet of modernity, a celestial body discovered not through ancient lore but through the deliberate application of science. Uranus symbolized the transition from a universe interpreted through stories to one understood through observation and analysis. It stood as a testament to the power of telescopes and the evolving techniques of astronomy to expand human horizons.

Despite its quiet demeanor, Uranus began to yield its secrets as technology advanced. The advent of spectroscopy in the 19th century allowed scientists to analyze the light reflected by the planet, revealing its composition and hinting at the presence of methane in its atmosphere. In the 20th century, the development of powerful telescopes and space missions provided even greater insights. Observations revealed its faint ring system, its family of moons, and its unusual axial tilt, each discovery adding to its mystique.

The Voyager 2 spacecraft's flyby of Uranus in 1986 was a transformative moment in the study of the planet. As the first and only spacecraft to visit Uranus, Voyager 2 provided humanity's closest look at this distant world. It revealed a planet that was far more complex than its serene exterior suggested. Uranus's rings, moons, and magnetic field came into sharp focus, and its atmosphere was studied in unprecedented detail. Yet, even with this wealth of data, Voyager 2 left many questions unanswered, highlighting the need for future exploration.

Uranus's mythology is not confined to its historical associations or its scientific discoveries. The planet has come to symbolize the value of

Uranus

subtlety and nonconformity in the broader human imagination. Its tilted axis, quiet demeanor, and pale appearance stand as a metaphor for difference—a reminder that the most intriguing phenomena often resist easy categorization or conventional expectations. Uranus challenges the idea that prominence must be tied to spectacle, offering instead a vision of grace in understatement.

The discovery of Uranus and the subsequent study of its features also underscore the importance of persistence in exploration. For centuries, Uranus had been visible in the night sky, mistaken for a star by countless observers. Its motion was too slow to attract notice, its brightness too faint to invite deeper scrutiny. It took the curiosity and dedication of Herschel to recognize its significance, an achievement that speaks to the power of patience and the willingness to question the familiar.

Philosophically, Uranus serves as a reminder of the limits of human perception and the value of expanding those limits. Its discovery marked the first time a planet was identified with the aid of technology, a milestone that opened the door to further breakthroughs in astronomy. Uranus is a testament to the idea that the universe is far larger and more diverse than we can imagine, and that our tools, techniques, and perspectives must evolve to meet its challenges.

In the quiet revolutions of Uranus's discovery and study, we find echoes of humanity's broader quest to understand the cosmos. It is a planet that invites reflection on the nature of progress, the beauty of difference, and the endless potential of exploration. Uranus's story is one of quiet impact, a reminder that even the most understated celestial bodies can inspire wonder and reshape our understanding of the universe.

Uranus

Chapter 7: The Forgotten Voyager

The Voyager 2 spacecraft's brief visit to Uranus in 1986 marked a fleeting but transformative moment in the exploration of the outer solar system. Of all the major planets Voyager 2 encountered during its grand tour, Uranus remains the least explored. The encounter was brief, the data limited by the technology of the time, and the public's attention overshadowed by the dramatic images of Jupiter and Saturn that preceded it. Yet, the flyby offered invaluable insights into Uranus and raised questions that continue to shape planetary science.

Launched in 1977 as part of NASA's Voyager program, Voyager 2 was designed to take advantage of a rare planetary alignment that allowed a single spacecraft to visit Jupiter, Saturn, Uranus, and Neptune. Its mission was to conduct flybys of these planets, capturing data and images that would transform humanity's understanding of the solar system. By the time Voyager 2 reached Uranus, it was the third planet in the spacecraft's itinerary, and its systems had already weathered nearly a decade of operation.

Voyager 2 approached Uranus on January 24, 1986, passing within 81,500 kilometers (50,600 miles) of the planet's cloud tops. For a spacecraft traveling at immense speeds, this close encounter lasted only hours, with months of preparatory work distilled into a rapid sequence of observations. During its flyby, Voyager 2 captured images, measured magnetic fields, and analyzed the composition of Uranus's atmosphere, rings, and moons. It provided the first detailed views of a planet that, until then, had been little more than a pale blue dot in distant telescopes.

The most striking revelation from Voyager 2's flyby was Uranus's **atmosphere**. Long considered featureless, Uranus was revealed to be a world of muted complexity. Voyager's cameras detected subtle bands and occasional bright spots in its cloud tops, evidence of limited atmospheric activity. The data confirmed that the atmosphere was composed primarily of hydrogen and helium, with methane responsible for its pale blue hue. Yet the lack of visible storms or dynamic weather patterns remained puzzling, likely tied to Uranus's unusually low internal heat.

Uranus

Voyager 2 also confirmed the existence of Uranus's **ring system**, which had been discovered nearly a decade earlier from Earth-based observations. The spacecraft revealed that the rings were composed of narrow, dark bands, starkly different from the bright, icy rings of Saturn. Voyager's images showed intricate details, including gaps and structures influenced by shepherd moons, and hinted at the dynamic processes shaping the rings over time.

Perhaps the most unexpected discovery during the flyby was Uranus's **magnetic field**. Unlike the magnetic fields of Earth or Jupiter, which are aligned relatively closely with their rotational axes, Uranus's magnetic field was wildly tilted, offset by 59 degrees from the planet's axis. Even more surprisingly, the field was not centered within the planet but appeared to originate far from its core. This asymmetry created a chaotic and lopsided magnetosphere, with regions of intense activity and turbulence. Voyager's observations of Uranus's magnetic environment were unprecedented, offering a glimpse into the complex dynamics of planetary interiors and magnetic fields.

Voyager 2 also turned its instruments toward Uranus's **moons**, providing the first close-up images of these distant satellites. Titania, Oberon, Ariel, Umbriel, and Miranda were revealed as diverse and intriguing worlds, each with its own unique features. Ariel's bright canyons and smooth plains hinted at recent geological activity, while Miranda's fractured surface and towering cliffs suggested a history of violent upheaval. These glimpses of Uranus's moons highlighted the variety and dynamism of the planet's satellite system, raising questions about their formation and evolution.

Despite these discoveries, the limitations of Voyager 2's technology and mission design left many questions unanswered. The spacecraft's instruments were not optimized for Uranus, and the brief nature of the flyby meant that much of the planet remained unexplored. Only one hemisphere of Uranus was imaged in detail, and the spacecraft's inability to linger or return limited the depth of its observations. These constraints left gaps in the data, particularly about Uranus's interior structure, seasonal changes, and atmospheric dynamics.

The timing of Voyager 2's encounter also posed challenges. The flyby occurred during Uranus's solstice, when one pole was bathed in sunlight while the other lay in darkness. This unique geometry offered valuable insights into Uranus's extreme seasons but limited the ability to study the planet's equatorial regions or observe how its atmosphere and magnetosphere evolved over time.

In the decades since Voyager 2's visit, Uranus has remained a largely unstudied world. No additional spacecraft have been sent to the planet, leaving much of the data gathered during the flyby as the foundation for our understanding of Uranus. Ground-based telescopes and the Hubble Space Telescope have provided additional observations, revealing occasional storms and seasonal changes, but these glimpses are limited by distance and resolution.

The scientific community has repeatedly called for a dedicated mission to Uranus, recognizing the planet as a key to understanding the diversity of ice giants and the processes that shape planetary systems. Proposed missions have included orbiters equipped with advanced instruments to study Uranus's atmosphere, rings, moons, and magnetic field in unprecedented detail. Such a mission could answer fundamental questions about the planet's origin, internal structure, and potential habitability of its moons, transforming Uranus from an enigma to a world of discovery.

Voyager 2's brief encounter with Uranus, while limited, remains a testament to the power of exploration and the value of venturing into the unknown. It was a mission that revealed the unexpected, showing that even the quietest and most remote planets are filled with complexity and wonder. Uranus's magnetic dance, its faint rings, and its moons' diverse landscapes all speak to the richness of the solar system and the need to continue exploring its farthest reaches.

Philosophically, the forgotten status of Voyager 2's visit to Uranus is a reminder of the vastness of the cosmos and the difficulty of fully comprehending it. Uranus's quiet presence challenges the assumption that discovery must be tied to spectacle, offering instead a vision of subtlety and resilience. Its mysteries remain an invitation to look closer, to seek the stories hidden beneath the surface, and to embrace the questions that linger long after the spacecraft has moved on.

Chapter 8: The Value of Difference

Uranus, with its quiet demeanor and subtle mysteries, stands as a symbol of the power and importance of difference. In a solar system dominated by the dynamic storms of Jupiter, the grandeur of Saturn's rings, and the dazzling blue of Neptune, Uranus's muted appearance and unconventional characteristics might seem understated. Yet, it is precisely in its nonconformity that Uranus reveals profound truths about the diversity of planetary systems and the importance of embracing the unexpected.

The planet's most distinctive feature, its extreme axial tilt, immediately sets it apart from all others. Rotating on its side, Uranus challenges the archetype of the upright, spinning planet. This tilt gives rise to some of the most unusual seasonal patterns in the solar system, with its poles alternating between decades-long periods of continuous sunlight and complete darkness. Such a configuration defies expectations and invites questions about the forces that shaped Uranus's early history.

Theories about a massive collision with a protoplanetary body provide one explanation for this unusual orientation, but the broader implications are equally fascinating. Uranus reminds us that planetary formation is not a uniform process. Chaos and unpredictability are intrinsic to the universe's creativity, capable of producing worlds that defy norms yet function within their own unique systems of order.

Uranus's muted pale blue hue further differentiates it from its neighbor Neptune, whose deep azure storms contrast sharply with Uranus's seemingly calm exterior. This subdued appearance might initially seem to diminish Uranus's allure, but it instead reflects the diversity of atmospheric conditions that can exist even between planets of similar size and composition. Methane gives both planets their blue tones, but subtle differences in atmospheric hazes and internal heat reveal Uranus as a distinct entity with its own story to tell.

The magnetic field of Uranus offers another stark departure from convention. Unlike the relatively aligned and centered magnetic fields of Earth or Jupiter, Uranus's magnetic poles are wildly offset and tilted in relation to its rotational axis. The resulting magnetosphere is chaotic and

Uranus

asymmetric, creating an environment of swirling charged particles and irregular auroras. This peculiarity invites scientists to rethink the mechanisms that generate magnetic fields, expanding models to include conditions that diverge from the norm. Uranus's magnetic field is not merely an anomaly; it is a puzzle that challenges established frameworks and opens new avenues of exploration.

The planet's ring system, often overlooked compared to the luminous rings of Saturn, is a marvel of understated beauty. Narrow and dark, Uranus's rings reveal a dynamism and complexity that reward close observation. Their faint glow, shaped by the gravitational influences of shepherd moons, speaks to the intricate interplay of forces that govern planetary systems. The rings of Uranus remind us that beauty in the cosmos is not always bold or immediate; it often lies in the subtle and the unexpected.

The moons of Uranus continue this theme of divergence and individuality. From Miranda's fractured surface and towering cliffs to Ariel's bright canyons and Titania's shadowy plains, each moon tells a unique story of geological activity, impacts, and evolution. They are worlds of contrast, shaped by tidal forces and their own histories, yet united by their orbit around Uranus. These moons underscore the diversity of processes that can occur within a single planetary system, offering a microcosm of the broader variety found in the universe.

Philosophically, Uranus stands as a symbol of nonconformity and resilience. It defies expectations at every turn: its tilt, its pale hue, its magnetic field, its faint rings—all refuse to fit neatly into established categories. Uranus invites us to question our assumptions about what planets should be and to embrace the differences that make each world unique. In a cosmos that thrives on diversity, Uranus reminds us that there is no single template for existence, and that deviation from the norm often leads to the most profound discoveries.

Uranus's value also lies in its ability to expand our understanding of planetary systems beyond our own. As we study exoplanets orbiting distant stars, many of which are ice giants similar in size to Uranus, the lessons learned from this tilted rebel take on even greater significance. Uranus challenges us to consider how factors like axial tilt, magnetic asymmetry, and atmospheric composition shape the characteristics of planets across the galaxy. By embracing the distinctiveness of Uranus, we refine our models and deepen our understanding of the processes that govern planetary diversity.

The human perspective on Uranus is itself a reflection of our capacity to grow and adapt. When Voyager 2 flew past the planet in 1986, it captured images and data that transformed Uranus from a pale dot into a world of dynamic systems and complex histories. Yet even today, Uranus remains a planet of unanswered questions, its mysteries a testament to the limits of our current knowledge and the potential for future discovery.

To appreciate Uranus is to recognize the value of difference not as a deviation from the norm but as a source of inspiration and insight. The planet's quiet presence, its understated beauty, and its unconventional features remind us that the universe is not bound by human expectations. It thrives on diversity, on worlds that break the mold and reveal the endless creativity of cosmic processes.

Uranus's story is one of quiet revolutions and subtle impacts. It challenges us to look beyond the obvious, to find wonder in the unexpected, and to celebrate the uniqueness of every world. The tilted rebel of the solar system, Uranus stands as a reminder that difference is not a flaw but a strength, a quality that enriches the cosmos and expands our understanding of what is possible.

Conclusion: The Planet of Quiet Revolutions

Uranus, the tilted rebel of the solar system, is a planet defined by its quiet revolutions. It does not command attention with the dramatic storms of Jupiter or the luminous rings of Saturn, nor does it exhibit the vibrant azure turbulence of Neptune. Yet, Uranus's understated presence conceals a world of complexity and depth, a planet that challenges norms and invites us to rethink what we know about planetary systems. Its story is one of subtlety and resilience, a reminder that the most profound discoveries often emerge from the least expected places.

At the heart of Uranus's identity is its extreme axial tilt, a feature that sets it apart from every other planet in the solar system. Rotating on its side, Uranus experiences seasons that are unlike anything else we know—decades of unbroken sunlight followed by decades of darkness at each pole. This unconventional orientation, likely the result of a colossal collision in its distant past, is a testament to the chaotic forces that shaped the early solar system. It is also a metaphor for Uranus's broader role as a planet that defies expectations, a world that exists on its own terms.

Uranus's pale blue hue, soft and muted compared to Neptune's deeper tones, reflects its quiet nature. Yet this seemingly placid exterior masks an atmosphere of intrigue. The lack of visible storms and the planet's minimal internal heat challenge our understanding of atmospheric dynamics, raising questions about how Uranus retains and dissipates energy. These mysteries deepen when combined with the planet's wildly tilted and offset magnetic field—a feature so irregular that it transforms Uranus's magnetosphere into a chaotic and asymmetric environment.

The rings of Uranus, faint and narrow, offer another example of the planet's understated complexity. While they lack the brilliance of Saturn's rings, they are remarkable in their own right, shaped by the gravitational influence of shepherd moons and dynamic processes that maintain their sharp edges. These rings are a quiet reminder that even in the vastness of space, order can emerge from chaos, and beauty can be found in the most delicate structures.

The moons of Uranus add further layers to its story. From Miranda's fractured surface to Ariel's bright canyons and Titania's shadowy plains, each moon is a world of contrasts and histories, shaped by tidal forces and geological activity. These satellites are more than companions; they are a testament to the diversity and interconnectedness of Uranus's system, each contributing to the planet's character in unique ways.

Voyager 2's flyby in 1986 offered humanity its closest glimpse of Uranus, transforming it from a distant, pale dot into a world of subtle complexity. Yet, even this historic encounter left many questions unanswered. The data collected during the flyby revealed a planet of profound mysteries but provided only a fleeting snapshot of its nature. Uranus's seasons, its internal structure, and the full extent of its atmospheric and magnetic dynamics remain subjects of speculation, awaiting future missions to uncover their secrets.

Philosophically, Uranus stands as a symbol of nonconformity and quiet strength. Its tilted axis, muted appearance, and unconventional magnetic field remind us that difference is not a limitation but a source of richness and diversity. Uranus challenges the notion that beauty and significance must be bold or obvious, offering instead a vision of grace in subtlety and complexity in restraint.

Uranus also speaks to the value of persistence and curiosity. For centuries, it was visible in the night sky but mistaken for a star, its slow movement across the heavens too subtle to notice. It took the ingenuity of William Herschel and the development of the telescope to recognize Uranus for what it was—a new planet, the first to be discovered in modern times. This milestone expanded the boundaries of the solar system and demonstrated the potential of science to reveal the unknown.

In a broader context, Uranus represents the quiet revolutions that shape the universe. Its discovery marked a turning point in astronomy, challenging long-held beliefs about the structure and limits of the cosmos. Its unique features continue to push the boundaries of planetary science, offering insights into the diversity of ice giants and the processes that govern their formation and evolution.

Uranus

Uranus's story is far from complete. The planet's pale hue and serene appearance are only the surface of a world filled with unanswered questions and untapped potential. Future missions to Uranus, equipped with advanced instruments and the ability to linger in its orbit, will undoubtedly deepen our understanding of this enigmatic world. They will reveal not only the physical processes that shape Uranus but also the lessons it holds for understanding planetary systems across the galaxy.

As humanity continues to explore the solar system and beyond, Uranus serves as a reminder of the importance of looking beyond the obvious. Its quiet revolutions—the tilt of its axis, the complexity of its rings, the resilience of its moons—are a testament to the diversity and ingenuity of nature. Uranus invites us to celebrate difference, to embrace curiosity, and to marvel at the subtle elegance of a planet that defies convention.

In the end, Uranus is more than a distant world in the cold reaches of the solar system. It is a symbol of the unexpected, a planet that challenges us to expand our perspectives and appreciate the quiet revolutions that shape both the cosmos and our understanding of it.

Uranus

End Note: Uranus's Lessons for Explorers

Uranus, the tilted giant, offers more than a study of planetary dynamics or a lesson in the diversity of celestial bodies. It is a world that challenges our perceptions, rewards our curiosity, and encourages us to explore beyond the limits of what we think we know. Its quiet demeanor belies its capacity to teach us profound lessons—not only about the cosmos but about the nature of exploration itself.

The first lesson Uranus imparts is the value of persistence in discovery. For millennia, Uranus drifted across the sky, unnoticed as anything other than a faint star. Its slow movement and dim light concealed its true nature from even the most attentive observers. It took the dedication and ingenuity of William Herschel, armed with the revolutionary tool of the telescope, to recognize Uranus for what it was: a planet that expanded the boundaries of the solar system and reshaped humanity's understanding of the heavens.

This discovery reminds us that the unknown is often hidden in plain sight, waiting for the right tools, the right perspective, or the right question to bring it into focus. Uranus teaches us that exploration is not always about seeking the extraordinary but about noticing the subtle and re-examining the familiar.

Another lesson lies in Uranus's quiet complexity. Its pale blue hue and seemingly calm surface mask a world of extremes and contradictions. From its dramatic axial tilt to its lopsided magnetic field, Uranus defies expectations and challenges traditional models of planetary behavior. This complexity is not loud or ostentatious; it requires careful observation and patience to uncover.

In this, Uranus reminds us that discovery is not always immediate or obvious. The most intriguing questions often lie beneath the surface, hidden within layers of data and requiring persistence to unravel. This is a lesson that extends beyond astronomy, speaking to the nature of inquiry in all fields. The greatest breakthroughs often come not from

Uranus

dramatic revelations but from quiet, careful work that brings hidden truths to light.

Uranus's unconventional characteristics also highlight the importance of diversity in the cosmos. Its extreme tilt, muted storms, faint rings, and dynamic moons demonstrate that there is no single template for what a planet should be. The differences between Uranus and its neighbors—Jupiter, Saturn, and Neptune—offer a richer understanding of the processes that shape planetary systems. Each world contributes something unique to the tapestry of the solar system, and Uranus's distinctiveness enhances, rather than diminishes, its value.

This is a powerful metaphor for human exploration and understanding. The diversity of Uranus and its system reminds us that difference is not a limitation but a strength. It is through encountering the unfamiliar, the unexpected, and the unconventional that we expand our knowledge and broaden our perspectives.

Uranus also serves as a case study for the limits of current technology and the promise of future exploration. The Voyager 2 spacecraft provided a brief but transformative glimpse of Uranus, revealing its rings, moons, and atmospheric dynamics. Yet the limitations of the mission—its brief flyby and the technology of its time—left many questions unanswered. What lies beneath Uranus's pale clouds? How does its magnetic field interact with its moons and the solar wind? What secrets are held by its faint rings and enigmatic satellites?

These questions remain a challenge to future explorers, and they underscore the importance of continued investment in planetary science. A dedicated mission to Uranus, equipped with modern instruments and the ability to linger in orbit, could unlock the answers to these mysteries and provide new insights into the nature of ice giants, both in our solar system and beyond. Such a mission would also advance our understanding of exoplanets, many of which share similarities with Uranus in size and composition.

Philosophically, Uranus speaks to the human capacity for wonder and resilience. It is a planet that requires us to look deeper, to ask harder questions, and to embrace the unknown with curiosity and humility. Its quiet revolutions and subtle complexities remind us that the universe is

Uranus

vast and varied, filled with phenomena that challenge our assumptions and expand our understanding.

Uranus's lessons are not limited to the scientific realm. They extend to the way we approach life and knowledge. The planet's unique characteristics remind us to value difference and to seek meaning in the unconventional. Its persistence in the sky, unnoticed for so long, teaches us the importance of patience and perseverance. Its quiet beauty encourages us to find elegance not only in the bold and dramatic but in the understated and the subtle.

In the end, Uranus is a planet that invites us to reconsider our place in the cosmos. It challenges us to think beyond the familiar, to celebrate diversity, and to continue exploring with an open mind and a sense of wonder. As humanity looks to the future, Uranus stands as a symbol of the unknown and the promise it holds—a reminder that the greatest discoveries often come from venturing beyond the boundaries of what we think we know.

Uranus, the tilted rebel, is not just a planet; it is a lesson in exploration, difference, and quiet resilience. It is a world that continues to teach us, inspiring us to look deeper, think broader, and dream bigger.

www.ingramcontent.com/pod-product-compliance
Lightning Source LLC
Chambersburg PA
CBHW070944220526
45469CB00007B/2513